Newton & Descartes's Day at the Beach

AF893532

MATH MUSICALS

www.MathMusicals.com

Volume 1

by
Anne Lazo
and
Michael Wiskar

BIG IDEAS LEARNING®
Erie, Pennsylvania
BigIdeasLearning.com

Cartoon Illustrations: MoreFrames Animation
Photo of Michael Wiskar provided by Greg Blue *www.gregblue.com*

Stockphoto credits:
MimaCZ/iStock/Getty Images Plus
dimonspace/iStock/Getty Images Plus
exxorian/DigitalVision Vectors/Getty Images
Vik_Y/iStock/Getty Images Plus
wissanu99/iStock/Getty Images Plus
Mjak/iStock/Getty Images Plus

Printed in the United States.

ISBN 13: 978-1-63598-900-7

123456789–22 21 20 19 18

Newton *Descartes*

Contents

The authors of *Big Ideas Math* understand the importance of storytelling and music in learning. Storytelling puts math in real-life contexts, and music helps children remember. In collaboration with storyteller Anne Lazo and musician Michael Wiskar, the authors of *Big Ideas Math* are excited to offer *Math Musicals*—engaging musicals that bring mathematics to life!

In this fictional world, students can interact with many different characters, including Newton and Descartes! Students will often wonder—What adventures will Newton and Descartes have today? What will they learn? What will they do next? Using *Math Musicals* will not only engage students in the mathematics, but will help improve their literacy skills as well.

Math and literacy skills are critically important in this world. Bringing math and literacy together through music is the perfect way to instill a lifelong love of learning, encouraging young minds to love math. At Big Ideas Learning, we believe providing the opportunity for students to interact with and learn the math through storytelling and music is just one more way to make learning visible.

About the Authors

Anne Lazo is an international performing arts educator, theatrical director, and author and composer of children's musical educational books.

Anne graduated from the University of Southern California and has over 20 years experience teaching and managing inspirational arts programs and productions for students, as well as professional actors, musicians, and dancers.

Anne worked in Japan for 15 years as an accomplished musical theater director, acting coach, choreographer, script writer, director of educational programs, and motivational speaker for arts advocacy. She has worked for Disney's World of English (World Family English), NHK Television and Radio, and owned her own production company, where she produced and directed musicals for schools and companies. In the United States, Anne was the Director of Education for the New West Symphony, where she developed and implemented many community outreach programs for all ages throughout the Los Angeles area.

Anne currently lives with her husband in Spain and is the Creator and Founder of My Child and Me - English©. Anne and her theatrical team of actors and teachers provide ESL educational shows, musical storybooks, and classes for young children.

Michael (Mike) Wiskar is a Leo-nominated composer, songwriter, independent music producer, vocalist, and multi-instrumentalist.

A graduate of the Music Program at MacEwan University as well as Langara's Digital Music Production Program, Mike began his music career in his teens as a performer in rock, country, and jazz bands before working as a songwriter for various artists in Quebec. There he worked closely with Myles Goodwyn of legendary Canadian rock band April Wine, among other top-level songwriters and producers.

Mike has since written and produced songs and music scores for theater, short films, full length and series documentaries, choirs, contemporary artists, and numerous television series (Operation Vacation, Psycho Kitty, Men's Fashion Insider, Kitty 911, and Favorite Places, to name just a few).

Math Musicals Stories and Topics

Grade K

Newton & Descartes's Coolest, Rockin' Day Ever

- *Fish Crackers* (Counting to 10)
- *Cool Cats & Rockin' Dogs* (Partner Numbers to 5)
- *Big Bulldog Bob* (Counting to 20)
- *JAM Session* (Naming Two- and Three-Dimensional Shapes)
- *Best Friends (I Get You and You Get Me)*

JAM Session

100 Waves

Grade 1

Newton & Descartes's Day at the Beach

- *Seashells* (Addition to 10)
- *Cora's Home* (Subtraction to 10)
- *100 Waves* (Counting to 100 by 10s)
- *Racing the Clock* (Telling Time)
- *Best Friends (I Get You and You Get Me)*

Grade 2

Newton & Descartes's Four-Legged Fun

- *Four-Legged Games* (Two-Digit Addition)
- *Newton & Descartes's Turn* (Two-Digit Addition and Subtraction)
- *Pet Airlines* (Counting to 1,000 by 100s)
- *At the Airport* (Money)
- *Best Friends (I Get You and You Get Me)*

At the Airport

Grade 3

Newton & Descartes's Night in Madrid

- *Tiny Tapas* (Multiplying by 3)
- *Flamenco Mice* (2s, 3s, 4s, and 5s Multiplication Facts)
- *Where is the Park?* (Fractions)
- *Fútbol Fun* (Perimeter)
- *Best Friends (I Get You and You Get Me)*

Fútbol Fun

Newton Leads the Way

Grade 4

Newton & Descartes's Skate Park Adventure

- *Jump Start* (Estimating Sums and Products)
- *The Amphibian* (Divisibility Rules for 2, 3, and 5)
- *The Catwalk* (Adding and Subtracting Fractions with Like Denominators)
- *Newton Leads the Way* (Types of Measurement)
- *Best Friends (I Get You and You Get Me)*

Grade 5

Newton & Descartes's Pet Center Adventure

- *Happy Paws Club* (The Language of Decimals)
- *The Cat's Eye Club* (Multiplying Decimals and Whole Numbers)
- *Bark Like a Dog* (Adding Fractions with Unlike Denominators)
- *Friends* (Volume of a Rectangular Prism)
- *Best Friends (I Get You and You Get Me)*

Friends

Accompanying Products

Visit *www.MathMusicals.com* for the following products.

Songs

Math Musicals songs are available for use throughout the year. Play the songs from *Math Musicals* to not only help students learn concepts, but to help them review all year long as well!

Animated Videos

The *Math Musicals* songs and storybooks come to life in these engaging animations filled with vivid and fun images that your students are sure to enjoy. Join Newton and Descartes on adventures around the world! From rocking out at a jam party to flying high in an airplane, Newton and Descartes engage students in mathematics in fun and innovative ways!

Sheet Music

Each *Math Musical* song comes with its own sheet music. Students and teachers can engage in learning by playing along with songs from *Math Musicals* on a piano or a guitar! Teachers have the ability to truly mix the teaching of math and music with this easy-to-read sheet music.

Differentiated Rich Math Tasks

Differentiated Rich Math Tasks are a fun and intriguing way for students to interact with mathematics. These leveled worksheets challenge students to stretch their knowledge and dig deeper! Each task has three different levels so teachers can easily differentiate to meet the needs of every student.

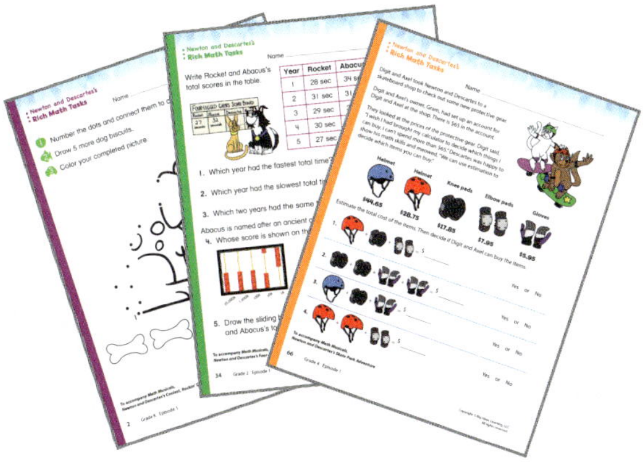

Puppets and Plush Toys

You can use the Newton and Descartes puppets and plush toys to make learning fun and to help students talk about math!

Newton & Descartes's Day at the Beach

by
Anne Lazo and Michael Wiskar

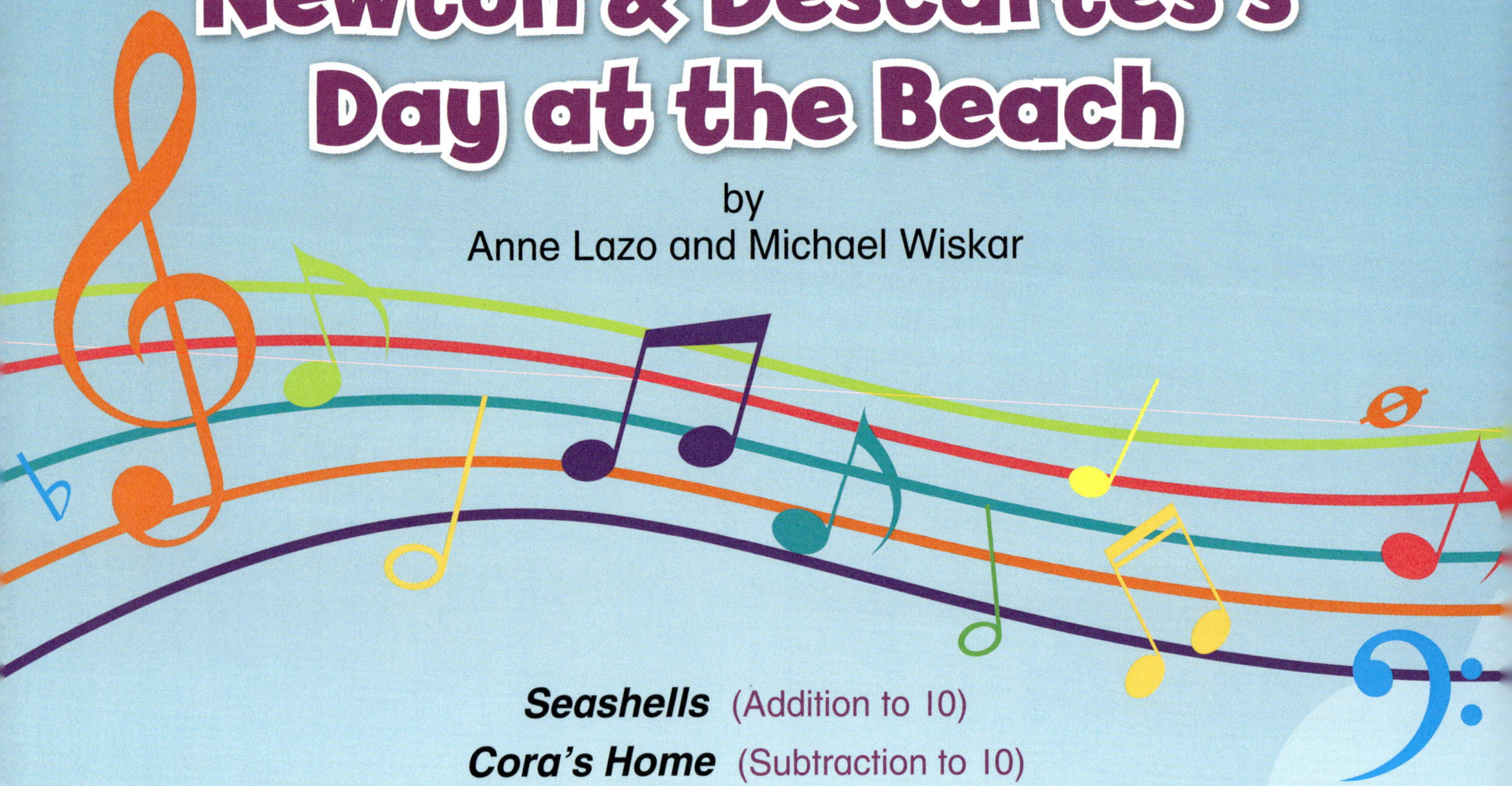

Seashells (Addition to 10)

Cora's Home (Subtraction to 10)

100 Waves (Counting to 100 by 10s)

Racing the Clock (Telling Time)

SEASHELLS

by Anne Lazo

It was a hot summer day when Miss Polly decided to go to the beach with her cute and clever pets, Newton and Descartes. She packed her picnic lunch and called for them to get into the car. "Newton, Descartes, please get into the car. We are going now!"

"I wonder where we are going today?" said Newton eagerly.

"I hope it's somewhere fun," replied Descartes.

As they ran into the car, they could see that Miss Polly had put a beach towel, an umbrella, a small chair, and a picnic basket on the back seat.

"Oh boy, oh boy," barked Newton. "We are going to the beach! Going to the beach, going to the beach. I love going to the beach!"

Descartes was not happy about going to the beach. He really does not like to swim and he certainly did not want to get full of sand and dirt. "I just want to rest at home today. Why do we have to go to a hot beach?" moaned Descartes.

"We get to play in the sand and swim in the sea," said Newton, as he stuck his head out of the car window and panted with excitement.

As Miss Polly drove out of the driveway, she told them that it was **10:00** in the morning and they had to return home by **3:00** in the afternoon. Miss Polly had invited friends over to the house for dinner and she needed time to get ready. They must not arrive home late from the beach! Descartes just sat, slumped on the back seat of the car, imagining how hot the sand would be on his delicate paws. He moaned the whole way there.

According to the clock in the car, **the big hand was on the 12 and the little hand was on the 11** so Descartes knew **it was 11:00** in the morning when they reached the beach. He yawned and sighed as he thought about how **their 3:00 deadline was a long 4 hours away**.

Miss Polly had barely opened the car door when Newton darted out from the back seat and ran onto the sand to explore. He likes to chase seagulls and run in and out of the waves.

“Don't run too far, Newton. Make sure I can see where you are!” said Miss Polly.

She set up her umbrella and chair and laid down her beach towel. Descartes stayed in the shelter of the umbrella and watched Newton make a spectacle of himself in the hot sun.

"Descartes, look!" yelped Newton as he dragged a seashell onto the sand. "Isn't this the prettiest shell you've ever seen? Look how big it is! There are a lot more over here. Come on, help me find some more shells!"

Descartes was not interested in finding seashells until he saw that one of the shells was moving. He decided that it was worth investigating to see what was inside. Descartes walked slowly to the shell and poked his nose in the opening. SNAP! "Ouch!" meowed Descartes. "Something just pinched my nose and it hurts!" He hit the shell with his paw and broke off a piece.

"Hey, what do you think you are doing?" snapped Cora the Crab.

"Who are you and what are you doing in there?" asked Descartes.

"My name is Cora and I am trying to protect myself from silly animals like you. That is why I live inside a shell. I can't live here now because you broke it. I have to look for another home. Please help me find a new one …. umm …. what is your name?"

"My name is Descartes. My friend, Newton, and I are looking for seashells."

"Well … you can have *this* one now. I don't want it. Start digging," said Cora. "We have to find a nice, big shell for me to live in because I've grown so much this year! My new home has to be spacious and not too heavy," she declared.

Cora the Crab started moving her legs through the sand, looking for a new shell home.

Descartes felt badly about breaking her home so he decided to help her find a new one. He dug in the sand with his paws. He found small shells, medium shells, and slightly larger broken shells. He wondered if any of them would be big enough for Cora.

Newton darted out of the sea and ran towards Miss Polly with another BIG shell in his mouth. He wanted to show her what he found, but she had fallen asleep after eating a big picnic lunch. "Oh well," he thought. "I'll show Descartes all of my shells instead and see what he thinks!"

Newton never saw Cora the Crab as he dragged his shells towards Descartes. "Wow, you found a lot of BIG shells!" said Descartes, as Newton dropped them in front of him on the sand.

"Yes, it's amazing!" said Newton. "It looks like you found a lot of shells, too!"

"I could only find some small, medium, and slightly larger broken shells. They don't look as nice as yours!"

"That's okay. Let's **add them together** and see **how many we have in all**!" said Newton with delight.

Descartes tried to explain to him about Cora the Crab. He was hoping that Newton's shells would be big enough for her to live in. Newton was not paying attention to Descartes because he was too busy lining up all the shells and adding them together.

I See Seashells

by Michael Wiskar

I see seashells,
So many seashells on the beach.
I see seashells,
So many seashells here.
How many have you got?
Pick them up!
How many have you got?

I've got 1. You've got 1.
1 and 1 is 2.

I've got 2. You've got 1.
2 and 1 is 3.

I've got 2. You've got 2.
2 and 2 is 4.

I've got 2. You've got 3.
2 and 3 is 5.

How many have you got?
I see seashells,
So many seashells on the beach.
I see seashells,
So many seashells here.
How many have you got?
Pick them up!
How many have you got?

I've got 3. You've got 4.
3 and 4 is 7.

I've got 4. You've got 4.
4 and 4 is 8.

Now I've got 5. And you've got 4.
5 and 4 is 9.

Now I've got 5. And you've got 5.
5 and 5 is 10!

How many have you got?

I see seashells,
So many seashells on the beach.
I see seashells,
So many seashells here.
How many have you got?
Pick them up!
How many have you got?
How many have you got?
How many have you got?
Pick them up!

Cora's Home

by Anne Lazo

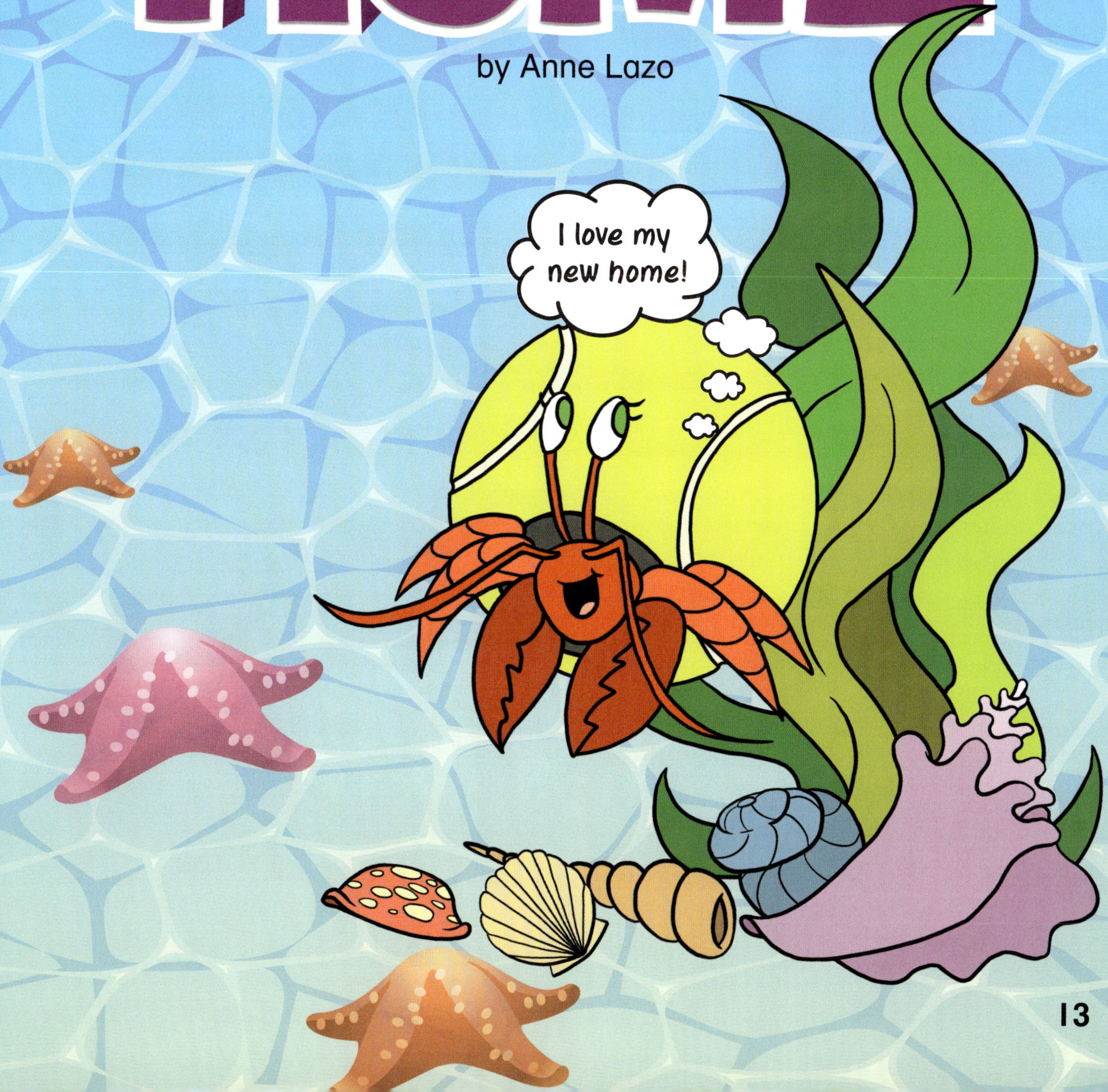

"I can't wait to show Miss Polly our shell collection! She is going to love them!" said Newton, as he admired all 10 shells.

"No, you can't take them from here," warned Descartes. "We need them for the crab!"

"Crab? What crab? What are you talking about?" asked Newton.

Suddenly, they saw Cora the Crab crawling on one of the big shells that Newton found. She could not believe her luck. "Wow! Thank you for bringing all of these fabulous homes to me! I can't wait to see which one fits!" said Cora eagerly.

"What?" yelped Newton. "These shells are mine. You can't have any of them!"

"Please share, Newton," said Descartes. "This is Cora and I broke her home. She used to live in this shell and now she has to find a new one. You found so many wonderful big shells. I am sure one of them will fit her and you can keep the rest!"

"A new home?" asked Newton.

"Yes, do you mind if I try them on for size?" pleaded Cora the Crab.

"Yes, of course I'll share with you. I found these shells in the sea! They are spacious, beautiful, and probably perfect for you!" Newton explained.

"Oh, thank you, Newton. That is so kind of you!" said Cora. "Finding a new home isn't easy anymore. Either the shells are broken and small, or don't exist at all. Out of these 10 shells, I am sure to find one that fits just right!"

Cora's New Home

by Michael Wiskar

10 seashells in a row.
Which one is going to be my house?
10 seashells, I know
I'm a little picky but I know what I like.
10 seashells, let's go!
Let's find a home for me that's right.

This seashell's a little too big.
10 minus 1 is 9.
These 2 here, a little too small.
9 minus 2 is 7.
These 2 here, hmmmm, I don't know.
7 minus 2 is 5.
Sorry, guys, I'm just not really feeling it.

5 seashells in a row.
Which one is going to be my house?
5 seashells I know
I'm a little picky but I know what I like.
5 seashells, let's go!
Let's find a home for me that's right.

This 1 here weighs too much.
5 minus 1 is 4.
These 2 here, not really my style.
4 minus 2 is 2.
This 1 here, no windows at all.
2 minus 1 is 1.
Sorry, guys, and this last 1
. . . it's just kind of ugly.

10 minus 1 is 9.
9 minus 2 is 7.
7 minus 2 is 5.
5 minus 1 is 4.
4 minus 2 is 2.
2 minus 1 is 1.

So what else have you got?

Newton and Descartes did not know what to do. None of the shell homes they found seemed to fit Cora the Crab just right. Cora was being very picky because she had to carry the shell around wherever she went. It had to be perfect!

Newton was looking around some rocks for more shells when he came upon a tennis ball that had been left on the beach. It was torn and worn, but it looked big enough to fit Cora comfortably.

"Look what I found, Cora," said Newton as he picked up the tennis ball in his mouth and gently dropped it down where she could easily crawl in.

"Wow, I've never seen a round shell like this before!" exclaimed Cora. "It looks wonderful!" She stepped inside the tennis ball and saw that it did, indeed, fit perfectly. Cora stretched out her claws and felt the inside of the ball. "It's so soft in here and look … it even has a great ocean view!"

The wind started to blow very strong and forced the tennis ball to roll around in the sand. But it did not seem to bother Cora.

"I am so excited! I can move around easily and quickly. I won't have to carry a heavy home on my back anymore! Thank you, both, very much!" she said, quite delighted with her choice.

"That is not a shell!" cried Descartes, as he watched Cora roll around.

"Now look what you've done. That is not a shell, Newton! She can't live in a tennis ball!" complained Descartes.

Before Descartes could explain to Cora that her new home was not a shell, a young boy picked up the tennis ball and ran over to Newton.

“Mommy, look, a ball and a cute doggy! Hello, fella. Come here, doggy, fetch this!” said the little boy to Newton.

The young boy had no idea that Cora was inside the tennis ball when he picked it up and threw it into the sea. He just wanted Newton to fetch the ball and bring it back to him.

It seemed like slow motion as the ball flew into the air. Newton and Descartes barked and meowed, “NOOOO! STOP!” But, to the boy, their cries sounded like animal gibberish, just barking and meowing. The tennis ball went flying into the air and out to sea. The waves swiftly took Cora, in the tennis ball, farther and farther out to sea.

100 WAVES

by Anne Lazo

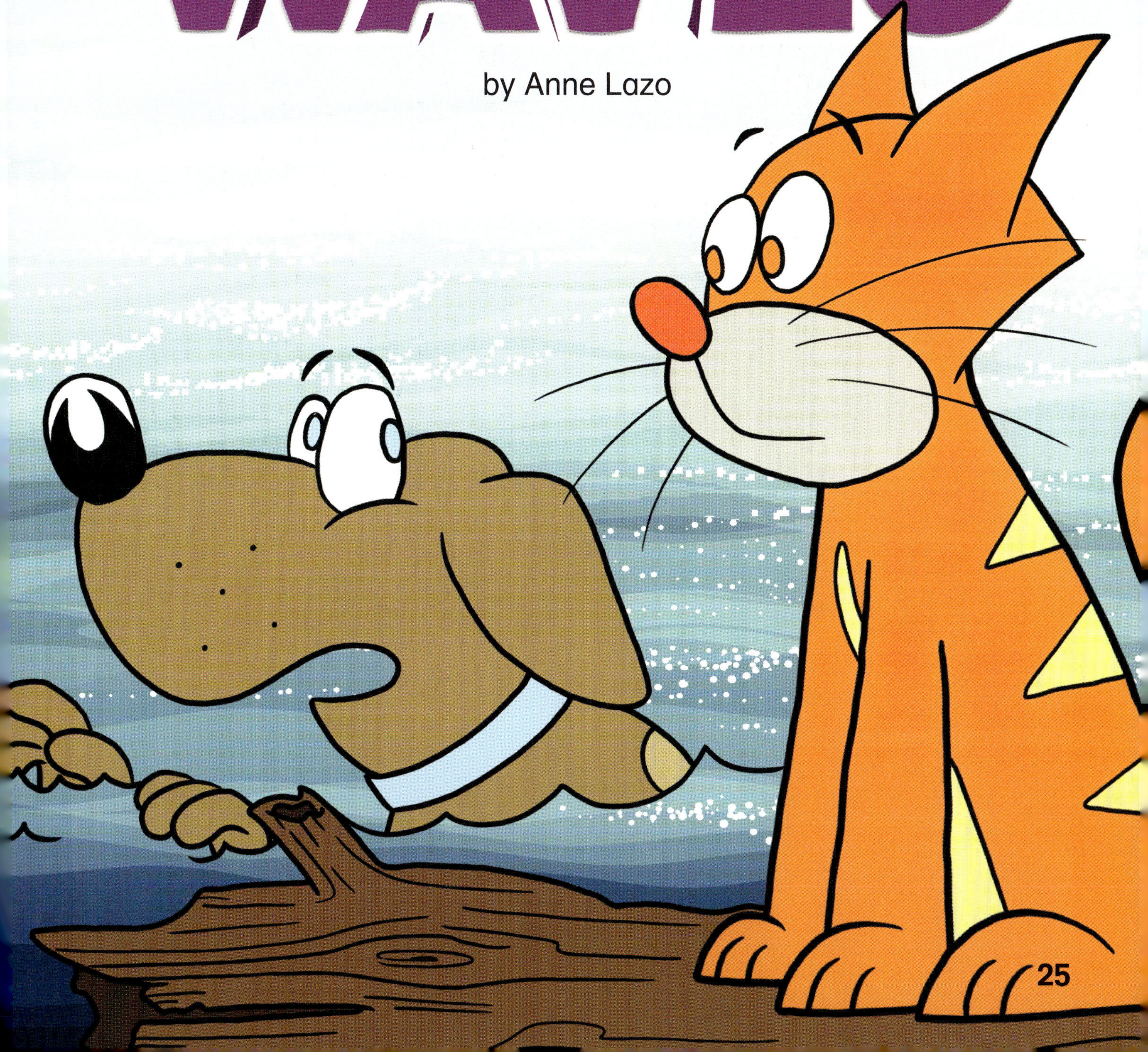

Newton and Descartes could not see where Cora the Crab had landed in the sea. They both started to worry. “We have to do something. Go swim after Cora, Newton!” said Descartes in a panic.

“I can’t swim in rough waves!” declared Newton.

“Listen, cats don’t swim in the sea, and even if I did, I would never be able to catch up with her!”

The young boy called out to Newton again and said, “Go get it, doggy. Go get the ball!” Newton could see that there were many rough waves. There were at least 100 waves! He knew he had to swim after Cora and bring her back to shore safely!

“You can do it, Newton. You can swim through 100 waves!” said Descartes.

Newton jumped into the sea and started to doggy paddle his way over the rough waves. Descartes decided to help his brave friend by jumping onto a piece of driftwood to ride the waves with Newton. Together, they would save Cora the Crab!

The young boy was so excited that Newton was retrieving the ball! "Mommy, look at the doggy go! Go, doggy, go! He's getting the ball!"

100 Waves

by Michael Wiskar

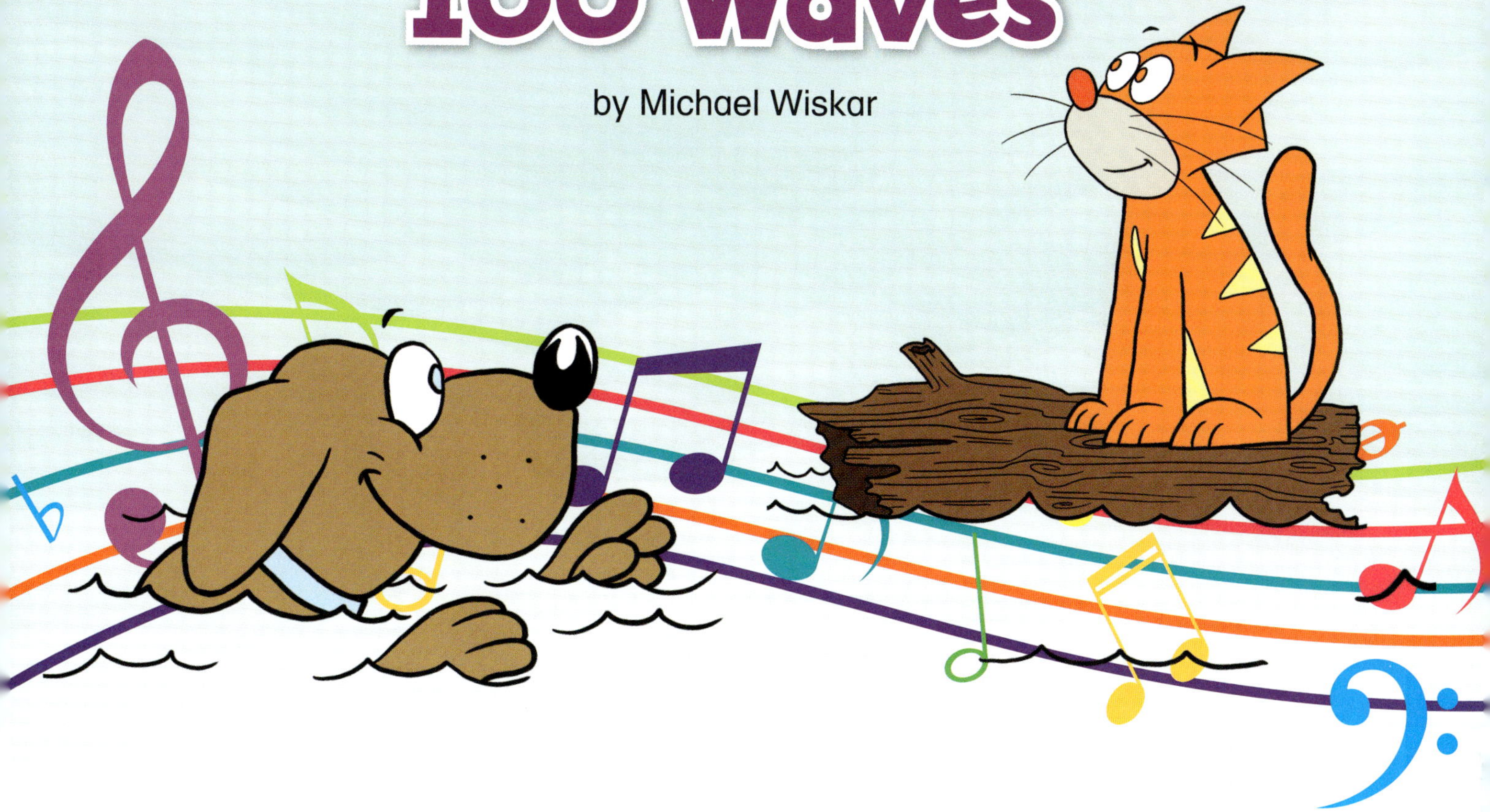

Here we go swimming in the waves,
100 waves coming our way.
Don't want to be swimming here all day.
Got to be home before it gets too late.

They go
Splish splash, splish splash,
splish splash,
And the waves keep coming.
They go
Splish splash, splish splash,
splish splash,
And the waves keep coming.

10, 20, 30
And the waves keep coming,
40, 50, 60
And the waves keep coming,
70, 80, 90
100 waves!

Here we go swimming in the waves,
100 waves coming our way.
Don't want to be swimming here all day.
Got to be home before it gets too late.

They go
Splish splash, splish splash,
splish splash,
And the waves keep coming.
They go
Splish splash, splish splash,
splish splash,
And the waves keep coming.

10, 20, 30
And the waves keep coming,
40, 50, 60
And the waves keep coming,
70, 80, 90
100 waves!

Newton and Descartes made it over the 100 waves! They were exhausted and sad because they could not find Cora anywhere. Wet and cold, Newton pulled Descartes on the piece of driftwood to safety. They found themselves on a tiny island that was just across the sea from where Miss Polly was still sleeping. They shook off the water and rested on the sand while the Sun's rays dried and warmed them.

Newton was panting hard, trying to catch his breath, when suddenly the tennis ball slowly rolled from the sea to his feet.

"Wow, what a house!" declared Cora the Crab as she poked her head out of her new home. "That was an awesome ride! You guys really know how to pick a house!"

"Cora, Cora, is that you?" yelped Newton. "We tried to rescue you, but we couldn't find you in all of those waves!"

"Thank goodness you are alright!" meowed Descartes, relieved.

"Wow, you both swam all that way to rescue me? That is so nice of you! True friends! But, you know, this home floats very well on top of water. I didn't go under once. I'm also a very good swimmer! When I get tired of walking on the sand, I move into the sea to find food," said Cora.

"What? You mean we went through all of that for nothing? Why didn't you tell us you could swim?" barked Newton.

"Please don't be angry. After all, you never asked!" said Cora with a smile. "I am glad that we all ended up on the island together. It's so beautiful here, and now we can all play together!"

"Yes, it is beautiful here," said Descartes, as he looked at the palm trees and the crystal blue water.

“I needed the exercise anyway,” said Newton. “You are right, Cora. It will be fun to explore the island together!”

They all laughed as they lay on the sand and took in the view.

RACING THE CLOCK

by Anne Lazo

As Newton, Descartes, and Cora rested on the island, they could hear the humming birds chirping away in the trees. "What are they chirping about?" asked Newton.

"I think they are telling their friends where they can find food in the trees," meowed Descartes, as he thought about food himself. He looked up at the birds and said, "I have to admit, I am pretty hungry for lunch right now!"

"Don't get any bright ideas!" said Cora jokingly. "We are guests on this island. You can't go chasing and eating the birds here! Don't look at me like that. You aren't eating me either! *I'm* not pet food!"

"Don't worry. After what you did to my nose, I wouldn't dream of trying!" teased Descartes. Both Cora and Descartes laughed at his bruised nose. Descartes was dreaming about fish, but confessed to Cora that he could not swim in the sea to catch one to eat!

"Let's explore the jungle and find something good to eat. It will be fun!" barked Newton, as his stomach was growling.

"I'll stay here," said Cora.

"I thought you wanted to play together?" asked Descartes, concerned.

"I do want to play but you guys are right; we should eat first. I can find plenty of plankton and algae right here, near the seashore," explained Cora.

"Alright, be careful, Cora," said Newton. "We will come back for you soon."

"If I hide in my unusual home, no one will even know I am here. I will be safe. Thanks!" said Cora proudly.

As Newton and Descartes started to walk into the jungle, they noticed it was thick with trees and strange plants. Newton started to sniff the ground for food, while Descartes climbed up a tree to see if he could find anything good to eat.

"Be careful, Descartes. Don't eat any leaves up there," said Newton, concerned that Descartes might get sick.

"I won't get sick," said Descartes, as he started to crunch on the bright green leaves of the tree.

Newton heard something in the distance and saw some bushes moving. He perked his head up and saw a turtle slowly crawling in one direction and then the other. "DESCARTES, A TURTLE!" called Newton.

"Of course I'm a turtle. Why don't you announce it to the world! Can't you see I'm lost and trying to make my way back to the sea? I walk very slowly and if you tell everyone that I am here, I am sure to be eaten by a big cat, or worse!" said the turtle.

The turtle looked at Newton and wondered who he was. "I've never seen an animal like you before in this jungle! You must be lost, too!" he exclaimed.

Newton explained to the frightened turtle how he met Cora the Crab and ended up on the island trying to save her life. He told the turtle that he was tired, hungry, and looking for something good to eat. While Newton was explaining what happened, Descartes leaped down from the tree to see what was going on.

"AAAH!" screamed the turtle. "I told you a cat would come and get me! HELP! HELP!"

"Calm down!" barked Newton. This is my friend, Descartes. He isn't going to hurt you. Look at him. He's a very *small* cat! Not a big jungle cat!"

"A cat is a cat!" cried the startled turtle.

"Excuse me, but I happen to be a very intelligent, sophisticated, and hip cat! Thank you very much!" proclaimed Descartes.

"I'll make a deal with both of you. I'll show you where you can find good food here, if you help me find my way back to the sea!" offered the turtle.

The turtle moved as quickly as his little legs could take him to a bush full of wild blueberries. He explained that the berries are '*sweet, full of antioxidants, and low in calories*!' The turtle thought that if the cat and the other *strange animal* ate the berries, they would leave him alone. After all, he did not believe Newton's story about finding a home for a crab and then trying to save her life!

"Wow, thank you!" said Newton to the turtle. "These berries look great! Why don't you start walking in that direction, towards the sea, and we will catch up with you when we are done eating."

"Fine," said the turtle. "But please call me, Mr. Turtle."

"Thank you, Mr. Turtle! We will only be a minute!" Newton barked cheerfully.

"You don't have to hurry," said Mr. Turtle. "I'm not going anywhere fast!"

Newton and Descartes enjoyed eating the blueberries so much that their faces turned blue as they gulped down as many as they could. When they were done eating, they ran up to Mr. Turtle and thanked him for introducing them to blueberries. They were yummy! Mr. Turtle was relieved that the cat and *strange animal* were full and happily followed them the rest of the way to the seashore.

Cora the Crab was pleased to see Newton and Descartes again! She could not see Mr. Turtle at first because he was some distance behind them. Cora wondered what they found to eat that made their faces turn blue.

“We made a new friend, Mr. Turtle, and he showed us some wild blueberries. We ate so many of them! They were delicious!” said Newton, still drooling from the thought.

“Mr. Turtle was lost, so we helped him find his way back to the seashore,” explained Descartes.

Just then, Mr. Turtle arrived at the shore and put his feet in the sea. "Ahhh!" he said, relieved. "I couldn't wait to do that!"

"A turtle! Oh dear!" Cora cried, as she hid inside the tennis ball.

"He won't hurt you, Cora. This is Mr. Turtle and he is very nice," purred Descartes.

"It's nice to meet you, Cora," said Mr. Turtle. "This cat and *strange animal* were so kind to help me when I was lost. Any friend of theirs is a friend of mine! By the way, that's a pretty neat shell you have there! I guess Newton was telling the truth about how he helped you find a home and then tried to rescue you at sea!"

Cora poked her head out of the tennis ball and looked at Mr. Turtle. "Yes, they were very kind to help me and now I have a home that fits just right! Gee … you have a shell, too. I guess we have a lot in common!" smiled Cora.

The wind started to pick up speed again and the waves were getting rougher. Newton looked out into the sea and realized that Miss Polly must be awake by now. She would notice they were gone and be frantic that they were missing! "We have to find our way back to the beach," Newton barked with concern. "How are we going to cross the sea? We can't swim over 100 waves again!"

"Don't worry," said Mr. Turtle. "I know where you can catch a boat to the mainland. It would take me too long to show you, so you must run and find the boat yourselves. Just follow the coastline for about a mile and you will see a dock with a boat. Try to jump on one that is leaving."

"Are you coming with us, Cora?" asked Descartes.

"It's so beautiful here and there's plenty of food for me. If I went with you now, I'd hold you back! You must hurry!" said Cora bravely. "I'm so grateful for everything that you have done for me! I'll always remember the kind pets who helped me find the perfect home!"

"Don't worry about her. We'll look after each other," said Mr. Turtle. "Now, run before the tide comes in!"

Newton and Descartes said goodbye to their sea friends and ran as fast as they could toward the boat dock. They finally found it, just where Mr. Turtle said it would be. There was a giant clock on the dock. **The big hand was on the 12 and the little hand was on the 2.**

"It looks like the next boat is leaving at **2:00**," said Descartes, as he wondered what time it was now.

A horn blew and the boat engine started to rumble.

"I guess **it is 2:00 right now**!" barked Newton. "Jump on, Descartes!"

The speed boat bounced up and down so hard on the waves that Descartes started to feel sick. "I told you, you shouldn't have eaten the leaves from that tree!" Newton said with a grin on his face.

The speed boat finally stopped at another dock across the sea, but at a different beach. Descartes looked at the clock on the boat to check the time. **The big hand was on the 6 and the little hand was between the 2 and the 3.** "Oh no!" meowed Descartes. "It's **2:30** already. We have to hurry, Newton! We only have **30 minutes left!**"

"What should we do now?" asked Newton, as he sniffed around. "I don't think we are anywhere near our beach yet!"

“Look! There’s a beach glider getting ready to take off! It’s bound to take us closer to Miss Polly!” Descartes called out in a loud voice. He motioned for Newton to leap for the glider and grab on!

“I…I….I just don’t know about this, Descartes,” whined Newton as they started to fly. “I am afraid of HE…HE...HEIGHTS!”

The beach glider flew right over Miss Polly’s head, but she did not see them. She was busy putting the umbrella, small chair, and the picnic basket back in the car.

“We have to jump off now, Newton!” screamed Descartes.

“No … no … I can’t … I can’t!” yelped Newton. “Look at all of the surfers down there. We’re going to hit them!”

“When I say 3, we have to jump. Ready? 1-2-3! Jump, Newton! JUMP!”

Before they knew it, both Newton and Descartes landed on a surf board. Clinging on for dear life, they rode over the biggest wave they had ever seen! Boy, was it scary! Finally, the surf board carried them safely onto the sand, right where Miss Polly was waiting for them.

Racing the Clock!

by Michael Wiskar

We're racing the clock
Got to make it to the dock
While the hands on the clock go
Tick tock tick tock!

We're Racing the Clock
It's much too far to walk
And we're in a state of shock
Tick tock tick tock!

The big hand is on the 12
The little hand is on the 2
It's 2 o'clock and what do we do?
We've got one hour to make it through.

We're racing the clock
Got to make it to the dock
While the hands on the clock go
Tick tock tick tock!

We're racing the clock
It's much too far to walk
And we're in a state of shock
Tick tock tick tock!

When **the big hand is on the 6**
And **the little hand's between 2 and 3**
It's 2:30 but what does that mean?
We've only **30 minutes until it's 3**!

We're racing the clock
Got to make it to the dock
While the hands on the clock go
Tick tock tick tock!

We're racing the clock
It's much too far to walk
And we're in a state of shock
Tick tock tick tock!

When **the big hand's on the 12**
And **the little hand's on the 3**
It's 3 o'clock . . .
Time's up!

We're racing the clock
Got to make it to the dock
We're Racing the Clock
Tick tock tick tock!

We're racing the clock
We're in a state of shock
We're racing the clock
Tick tock tick tock!

“Where have you two been? Oh no … you are soaking wet! Have you been swimming all day?” asked Miss Polly, as she dried them off with the beach towel. They were so excited and relieved to see her that they jumped up and down with joy!

“Please, get into the car now. We are going home,” said Miss Polly. She was delighted that her cute and clever pets were happy and made it back on time!

As they ran into the car, Descartes looked at the clock. **The big hand was on the 12 and the little hand was on the 3. It was 3:00!** “We really made it on time. Amazing!”

Descartes took one last look at the sea. He thought about how he did not want to spend **4 hours** on the beach at first. Now, he was glad that he went! What an adventure they had! He would never forget his friends, Cora the Crab and Mr. Turtle, and how brave Newton was trying to save Cora from the 100 waves!

Miss Polly looked through the rearview mirror of the car and noticed something odd about Newton and Descartes. "Why are your faces blue?" she asked, confused.

"Descartes, what happened to your nose?"

Best Friends
(I Get You and You Get Me)

by Michael Wiskar

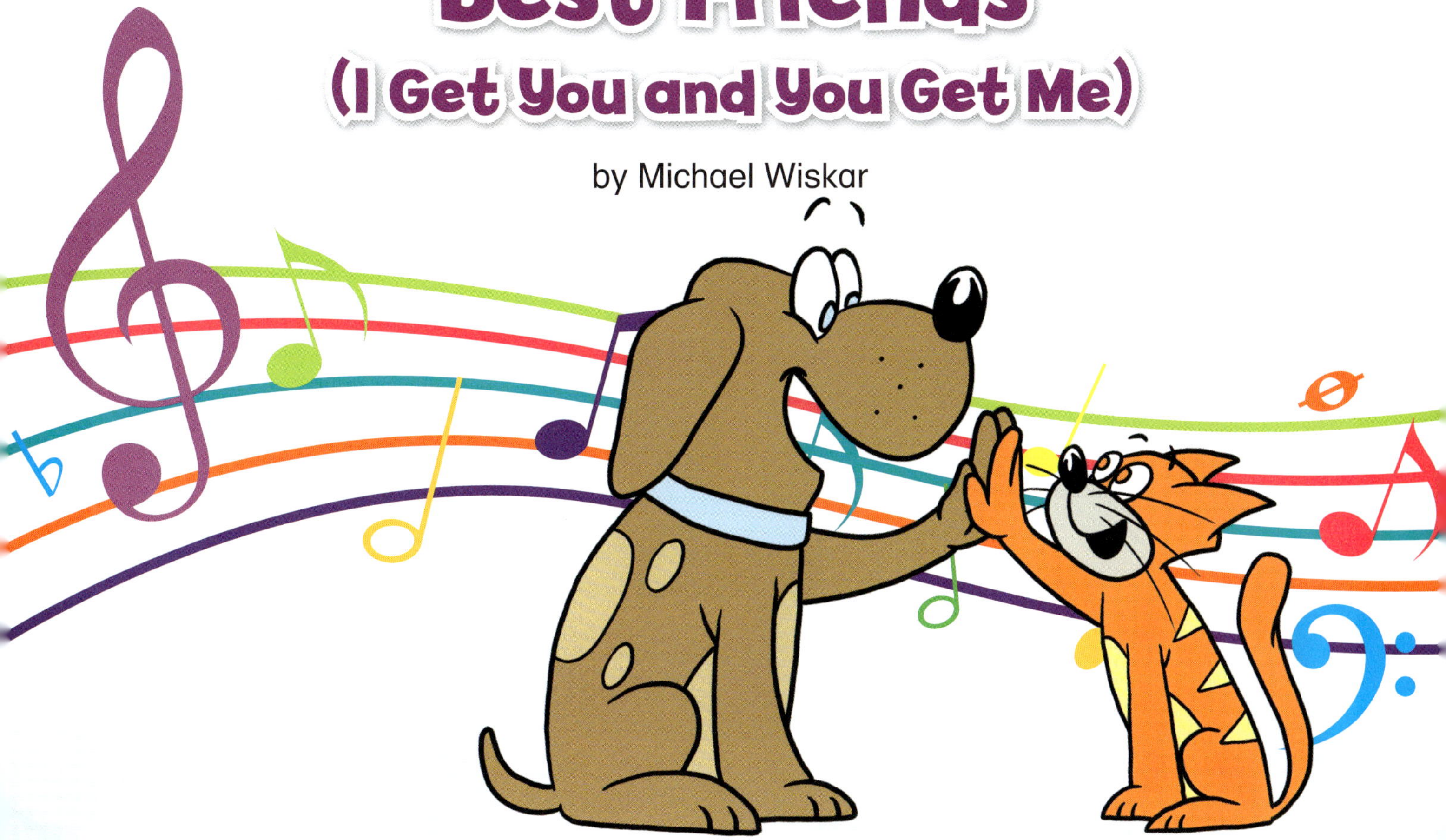

A wink, a nod, a sideways glance,
a little elbow to the ribs.
We speak a secret language and
no one else knows what it is.
You know me better than
I even know myself.
It's a mutual connection
between us and no one else.

You've got a friend,
Whenever you need a friend.
You've got a friend,
Whenever you need a friend.
I can't pretend
there's anywhere else I'd rather be,
'Cause I get you and you get me.

And even when there's nothing to do
we're never feeling bored.
We've had a thousand adventures
and we'll have a thousand more.
You make me laugh so hard
I feel like I could cry.
And we pick each other up
when we're together,
We don't even have to try.

You've got a friend,
Whenever you need a friend.
You've got a friend,
Whenever you need a friend.
I can't pretend
there's anywhere else I'd rather be,
'Cause you've got a friend in me.

'Cause I get you and you get me.
'Cause I get you and you get me.
We're a special kind of family.

And even if there's trouble brewing
we always know just what we're doing,
'Cause I've always got your back
and you've got mine.
And everything's fine!

You've got a friend,
Whenever you need a friend.
You've got a friend,
Whenever you need a friend.
I can't pretend
there's anywhere else I'd rather be,
'Cause I get you and you get me.

Whenever you need a friend.
('Cause I get you and you get me.)
Whenever you need a friend.
I can't pretend
there's anywhere else I'd rather be,
'Cause you've got a friend in me.

'Cause I get you and you get me.
(Yeah, I get you!)
'Cause I get you and you get me.
(I think you get me too.)
We're a special kind of family.

Let's sing it!

You've got a friend,
('Cause I get you and you get me.)
Whenever you need a friend.
You've got a friend,
('Cause I get you and you get me.)
Whenever you need a friend.
You've got a friend,
('Cause I get you and you get me.)
Whenever you need a friend.
We're a special kind of family.

You've got a friend,
('Cause I get you and you get me.)
Whenever you need a friend.
You've got a friend,
('Cause I get you and you get me.)
Whenever you need a friend.
You've got a friend,
('Cause I get you and you get me.)
Whenever you need a friend.
We're a special kind of family.

You've got a friend,
('Cause I get you and you get me.)
Whenever you need a friend.
You've got a friend,
('Cause I get you and you get me.)
Whenever you need a friend.
You've got a friend,
('Cause I get you and you get me.)
Whenever you need a friend.
We're a special kind of family.

I See Seashells

Words and Music by
Michael Wiskar

A E
39
So ma-ny sea-shells here. How ma-ny have you got? Pick them up! How ma-ny have you got?
44
I've got three. You've got four. Three and four is se - ven. I've got
49
four. You've got four. Four and four is eight. Now I've got
A E B A
53
five. And you've got four. Five and four is nine. Now I've got five. You've got five.
E
59
Five and five is ten. How ma-ny have you got? I See sea - shells, So ma-ny sea-shells
64
on the beach. I see sea - shells, So ma-ny sea-shells here. How ma-ny have you
A E
69
got? Pick them up! How ma - ny have you got? How ma - ny have you
B A E
73
got? How - ma - ny have you got? Pick them up!

Cora's New Home

Words and Music by
Michael Wiskar

Dmi
24
I'm a lit-tle pick-y but I know what I like. Five sea shells, let's go! Let's find a home for me that's right.
Cma7
F(♭5)
7fr.
27
This one here, weighs too much. Five mi-nus one is four.
Emi
Dmi
32
These two here, not real-ly my style. Four mi-nus two is two. This one here, no win-dows at all.
Cma7
36
Two mi-nus one is one. Sor-ry, guys. And this last one It's just kind of ugly.
F(♭5)
7fr.
Emi
40
Ten mi-nus one is nine. Nine mi-nus two is se-ven Se-ven mi-nus two is five. Five mi-nus one is four.
Dmi
Cma7
44
Four mi-nus two is two. Two mi-nus one is one. So what else have you got?

100 Waves

Words and Music by
Michael Wiskar

E7 A E7
28
waves keep com-ing, se-ven-ty, eigh-ty, nine-ty, one hun-dred waves!
A7 E
33
Here we go swim-ming in the waves. One hun-dred waves com-ing our way. Don't
A7 E7 A G
37
want to be swim-ming here all day. Got to be home be-fore it gets too late. They go
A E7 A
41
splish splash, splish splash, splish splash, And the waves keep com-ing. They go splish splash,
E7 A
46
splish splash, splish splash, And the waves keep com-ing. ten, twen-ty,
E7 A
51
thir-ty And the waves keep com-ing, for-ty, fif-ty, six-ty And the
E7 A E7 A
56
waves keep com-ing, se-ven-ty, eigh-ty, nine-ty, one hun-dred waves!

Racing the Clock

Words and Music by
Michael Wiskar

F♯mi E F♯mi/A Emi/G C♯7
28
And the little hand's between two and three
It's two thirty, but what does that mean?
We've on - ly thir - ty min - utes un -
F♯mi
33
til it's three! We're ra - cing the clock Got to make it to the dock While the hands on the clock go
Bmi F♯mi
37
tick tock tick tock! We're ra - cing the clock It's much too far to walk And we're in a state of shock.
Dma7 C♯ C♯mi D♯7 Dma7
41
tick tock tick tock!
When the big hand's on the twelve
And the little hand is on the three
G♯ 4fr. C♯aug F♯mi
47
It's three o'clock.
Time's up! Aaarg! We're ra - cing the clock Got to make it to the dock. We're
52
ra - cing the clock. tick tock tick tock! We're ra - cing the clock We're in a state of shock. We're
F♯mi
56
ra - cing the clock tick tock tick tock!

Best Friends (I Get You and You Get Me)

Words and Music by
Michael Wiskar

D/A F♯mi E A D/A
thing to do we're ne - ver feel - ing bored. We've had a thou - sand ad - ven - tures and we'll have
F♯mi E A/C♯ D F♯mi E
a thou - sand more. You make me laugh so hard I feel like I could cry.
A/C♯ D F♯mi E
And we pick each o - ther up when we're to - ge - ther, we don't ev - en have to try. You've got a
D.S. AL CODA
CODA
D/E A Emi7 A
you've got a friend in me. Yeah 'Cause I get you and you get me. 'Cause
Emi7 A Emi7 DMA9
9fr.
I get you and you get me. We're a spe - cial kind of fa - mi - ly.
F♯mi E G F♯mi
And ev - en if there's trou - ble brew - ing we al - ways know just what we're do - ing, 'Cause I've
B D/E E7
al - ways got your back and you've got mine. And ev - 'ry - thing's fine. You've got a
D.S. AL CODA

CODA
D/E A D/G A D/G
I get you and you get me. 'Cause I get you and you get me. need a friend. I can't pre-
When-ev-er you need a friend. When-ev-er you
F♯mi A/C♯ Bmi D/E A
tend there's an-y-where else I'd ra - ther be. 'Cause you've got a friend in me. 'Cause
Emi7 A Emi7 A Emi7
I get you and you get me. 'Cause I get you and you get me. We're a spe-cial kind of fa -
Yeah, I get you. I think you get me too.
DMA9 A Emi7 A
mi-ly. 'Cause I get you and you get me. 'Cause I get you and you
Let's sing it!
You've got a friend, when-ev-er you need a friend. You've got a friend, when-ev-er you
Emi7 A Emi7 Emi/D 1, 2. DMA9
get me. 'Cause I get you and you get me. We're a spe-cial kind of fa - mi-ly. 'Cause
need a friend. You've got a friend, when-ev-er you need a friend. We're a spe-cial kind of fa - mi-ly. You've got a
3. DMA9 A
mi - ly.